Rational Treasure

II

Rational Treasure

How a few simple tricks can change the way you think

By

Aryan Singh

Copyright © 2018 by Aryan Singh.
All rights reserved.
ISBN-13: 9781731509505

Dedication

This book is dedicated to my two young irrational brothers, Nirmay Singh (10yrs) and Alekay Singh (7yrs).

Table of Contents

Chapter 1: Introduction — 1
Chapter 2: No avail — 17
Chapter 3: Pulled in too deep — 29
Chapter 4: Reference of frames — 43
Chapter 5: Over your head — 55
Chapter 6: The idol of endowment — 69
Chapter 7: No change to spare — 75
Chapter 8: Working it out — 89
Chapter 9: Fifty shades of green — 109
Chapter 10: Conclusion — 121

Chapter 1: Introduction

Before I start, I want to introduce two (badly drawn) people. First, meet Hugh Mann.

Now, I just want to explain that Hugh is a human. It's so obvious that people forget it. I mean, when you think of someone, human isn't exactly the first word you associate with them. Now meet Ike O'nomist.

Ike is different from Hugh in that he's not human. As you might have figured out from his name, Ike is instead an economist.

Now Ike has two defining traits, which he thinks all humans have, too. He's incredibly greedy,

and incredibly rational.

However, over the course of this book, Ike and you will realize that humans are far from rational. In fact, humans are terrible at making rational decisions.

You're probably already as skeptical as Hugh. How can humans not be rational? We're the smartest species on the planet! If that's what you're thinking right now, try this simple question. A bat and a ball together cost $1.10. The bat costs a dollar more than the ball. How much does the ball cost?

Wasn't that easy! The ball is obviously 10 cents! Who would be so stupid that they can't figure that out!

You caught it, Ike. If the ball cost 10 cents, the bat would cost $1.10, and they would be $1.20 together. The correct answer is 5 cents. But most humans, including you, probably thought the ball was 10 cents. This is because a human doesn't take the time to think about questions like these, and the obvious answer quickly comes to their head. And if you already think humans are stupid, then wait until you read this book, because in these 10 chapters, I'll explain how humans use random numbers as evidence, work less with more money, and eschew actual facts for interesting stories.

Oh, right, I almost forgot about the map. A few weeks ago Ike bought a used briefcase from the thrift store, to save money, of course.

But recently he discovered that a map was hidden in the briefcase. A map that leads to a treasure beyond all wealth.

So Ike and Hugh decided to embark on a great adventure to find the treasure and become incredibly rich. In fact, you can come along too, and while embarking on this quest, discover the biases all humans have, and learn to avoid them and exploit them. So are you ready to travel to an unknown land, find clues to get to your goal, conquer multiple trials and challenges, and discover the lost treasure of Econ Island? Yes? Then you can go as soon as Hugh and Ike finish arguing about how to get to the island.

Chapter 2: No avail

Okay, this is starting to get old. Hugh and Ike have been arguing for hours over whether they should go to the island by plane or boat.

What's worse, they just keep repeating the same arguments! First Hugh talks about all the plane crashes that happened recently:

Then Ike says that there were many more boat crashes than plane crashes in that time but they didn't get on the news.

And they just go on and on.

But if you pay attention to this argument, it's pretty clear that Ike is right. All the statistics and concrete evidence prove that boats sink way more than planes crash. But Hugh still insists that planes are dangerous.

People do this all the time, and it's called the availability heuristic. Hugh just made an assumption, which isn't backed up by statistical evidence, but by an example that's easy to remember. And people fall prey to it all the time. One example is very well known, and surprisingly enough, has to do with eating laundry detergent.

Yes, the Tide PODS® epidemic, in which thousands of teens have been eating Tide PODS® for fun, is a prime example of the availability heuristic. Though the crisis itself wasn't caused by this bias, a lot of the news stories surrounding it were. For example, multiple people were fretting over the fact that in 2017, there were 12,299 poison control calls about kids eating detergent pods. That sounds like a scary number, but if you compare it to over 20,000 calls for kids drinking hand sanitizer, 17,000 for toothpaste eating, 16,000 for deodorant munching, and 13,000 for mouthwash chugging, it doesn't sound as big. In fact, you might be wondering why people aren't talking about the hand sanitizer epidemic, or why the toothpaste eaters aren't coming up on the news! This is all because of the availability heuristic. Teenagers eating Tide PODS® for an Internet challenge is more noticeable than little kids drinking hand sanitizer because they don't know any better (especially since teens post their laundry pod feast on social media).

Another current and well-known example of the availability heuristic is when Donald Trump, the president of the United States, posted a list of every single crime done by immigrants.

Ike's right to be frustrated. This is stupid! Trump made it look like immigrants commit many crimes, but imagine what would happen if he posted a list of every single crime ever done?

Do you see what I mean? Just because Trump is prejudiced against immigrants, he made a list, which irrationally bends the truth. Many things make something easier to remember or access, for example:

Tragedies are more memorable than happy endings.

Events gain more attention if they are captivating.

Availability depends on what's current.

How to avoid it

Suppose you're in an argument like the one Hugh and Ike are having now. If the other person shows you some anecdote or news story or anything like that, you know what to do.

No, you ask for statistical evidence.

For example, if somebody shows you news about recent bear attacks, you can ask them for statistics on how many bear attacks happened this year. Then, you could show them the number of car accidents in the same year (which outnumbered bear attacks). After all, we consider driving to be safe, so shouldn't hiking in bear habitats be safe too? If you use this method, the other person will not know how to retaliate.

How to exploit it

Of course, just because you're avoiding the availability heuristic, doesn't mean you can't also use it to your advantage. Imagine the same argument, but this time the statistics are working against you. Well then you can look for anecdotal evidence to support your claim, and purposefully ignore statistics. Your opponent, being as irrational as all other humans, won't even notice!

I can see where you're coming from, Hugh. It's a case where there's nothing separating the good guys from the bad. But this isn't good and bad; these are just human biases. You can avoid them, and you can use them to your advantage.

Chapter 3: Pulled in too deep

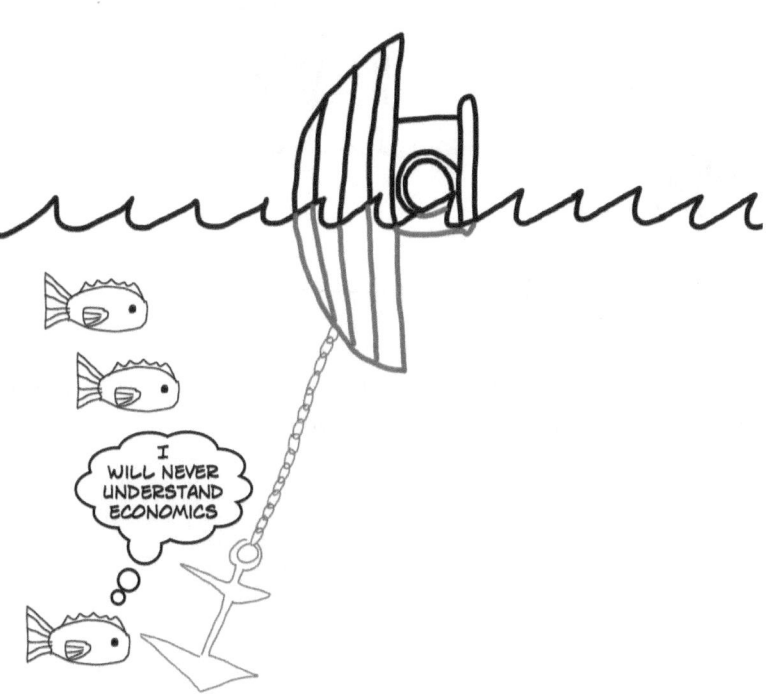

Well, it seems like Hugh won the argument, but their boat is missing an anchor. So Hugh and Ike have gone out to buy a new anchor for their boat, with $45 to spare. And lucky for us, it can be a learning experience.

Ike has been to this store before, so he knows that there are usually three different anchors on sale.

A wooden anchor for $15

An iron anchor for $30

And a sturdy titanium anchor for $45

Last time Ike was at this store, he chose the iron anchor, because it wasn't fragile like the wooden anchor, and it wasn't expensive like the titanium anchor. Most humans would make this choice, too.

Not so fast, Ike. You might think you won, but you haven't seen anything yet. Today, they have the same three anchors for sale, but they also have a fourth anchor. A gold-plated, diamond-studded, ultra-strength, anchor.

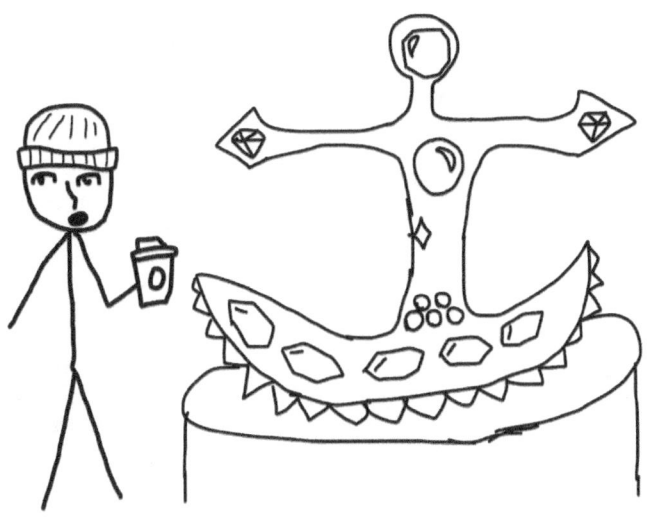

For $480.

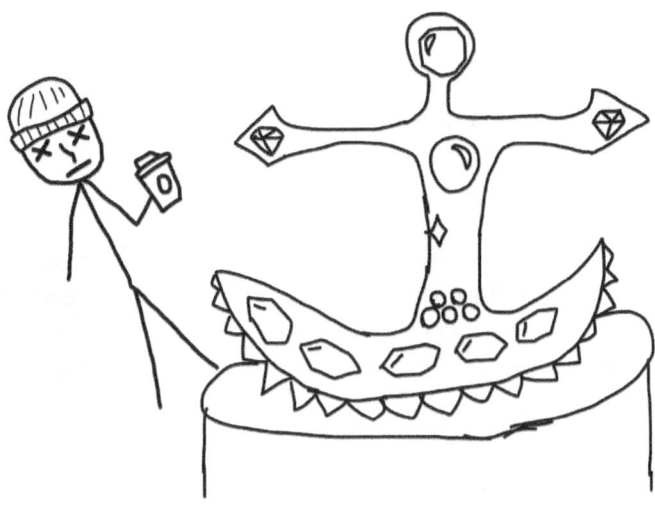

Now, let's see what's going on with Hugh and Ike.

Do you see what I meant when I said humans can't make good decisions? If that super-expensive anchor weren't there, Hugh would have bought the iron anchor for $30. But just because the storeowner decided to offer a super-expensive anchor that no one would buy, Hugh bought the titanium anchor for $45! But don't think Hugh is just an idiot.

Everybody is susceptible to this trap, even you! If you were in Hugh's place, you would have done the exact same thing! This is actually a psychological phenomenon, and behavioral economists call it, appropriately enough, anchoring. However, retailers figured this out before behavioral economists did. In fact, the storeowner might have put that anchor there intentionally, knowing nobody would buy it, so that people would buy the titanium anchor!

But first let's get to how anchoring works. When you see a number, your mind fixates on it, and you end up anchored to the number. If you're asked questions later on the subject of the first number, your answer will be based on the first number. The titanium anchor looked better in comparison to the expensive anchor, so Hugh was compelled to buy it.

Ike's right. It is absurd. But you do it every day. Not just when you're buying stuff, but whenever you're making a decision, there could be one stray piece of information secretly controlling what you do, even if it's completely unrelated to the decision you're making. At least that's what happens according to studies of adults performed by accomplished economists. But I decided to run an experiment myself to prove that it works on kids too. I asked multiple students from a sixth and seventh grade classroom: Did Bruce Lee make more than 300 movies, or less than 300 movies? After that question, I asked them: How many movies did Bruce Lee make? As you can tell, basically all of them knew that Bruce Lee made less than 300 movies. However, their guesses for the second question were around 150 or 200, because they were anchored to the number 300. So how many movies did Bruce Lee make? 18. This is because Bruce Lee died very young. In the sample some people knew that Bruce Lee died young; some had seen Bruce Lee movies. Those people were closer to the answer, but they were still manipulated by the anchor, which was put there on purpose to mislead people. When you have little information, anchoring works best. The more you know, the less the anchor can pull you in.

But of course, you can't have a good experiment without a control sample. So some of the students got a sheet that simply asked them how many movies Bruce Lee made. These students weren't pulled by the anchor, and gave answers that were closer to the correct answer. And some of the students got a sheet with a different anchor completely: Bruce Lee died very young. How many movies did he make? These students were still being anchored, but in a different way; they ended up getting closer to the answer.

But why does anchoring happen? System 1, your intuitive system, causes a lot of the biases in this book. But anchoring is an exception. It's a tool used by System 2, your analytical system. You anchor on a number so that you have a starting point to base your decision on. Take this example:

What year was George Washington elected president?

Even if you don't know the answer to this question, you might remember that America gained independence in 1776, so your answer should be a few years after that. But this helpful trick can also backfire, and your anchor might do you more harm than help.

How to avoid it

When a number is anchoring you, the anchor is usually irrelevant information, stuff that isn't related to the answer, stuff you don't really need. So, you should at least try to ignore the irrelevant information when making a decision. Just put it aside and never look at it again. Make sure your answer isn't swayed by that information.

But in some cases, like how Hugh and Ike bought anchors, you can be anchored by something that is relevant to the decision you're making. This usually happens when multiple options are shown. In that case, the best strategy is to judge each option individually. If Hugh had done that, he would have decided that the iron anchor is a well-rounded option for a good cost, instead of making the decision based on other anchors.

How to exploit it

Once you know how to avoid a bias, you can learn to exploit it, and use it to your advantage. In the case of anchoring, it is incredibly useful for getting your parents to do what you want.

Suppose you want a raise in your allowance. You can use anchoring to fool your parents into giving you a higher raise. Here's how, demonstrated by Hugh's son Hector:

As demonstrated, first ask for a super high allowance that you know your parents won't agree to. This is your anchor; the number you will use to sway them in your direction. Then ask for a raise that is a bit lower than the anchor, but still too high for your parents to agree with normally.

It worked! The only reason they agreed to that was because they fell for anchoring. Your parents might be smarter than you (which is why you should always listen to them), but nobody is perfectly rational, not even them. Try it on your parents. It should work. And you can use this technique for other things too. A few examples:

Chapter 4: Reference of frames

Despite their mishaps with the titanium anchor, Hugh and Ike made it to Econ Island, where the treasure is hidden. But before they start looking for the treasure, they're going to take a break at the carnival. But have no fear reader; you will get your money's worth. For we can continue even while they're taking their break. Looks like Hugh's at the tarot reader already.

Let's see what Hugh thinks of that fortune, shall we?

Well, he didn't like it very much. Now he's going to another tarot reader.

And what does he think of that one?

Apparently, Hugh hates funerals. But here's the thing: Both fortunes are the same. The only reason you wouldn't go to your family's funerals is if you were the first one in your family to die! So why does Hugh have such different reactions to them? One word:

Framing is the idea that something can seem better if it's presented in a certain way. For example, furniture that will last 10 years 90% of the time, sounds better than furniture that has a 10% chance of breaking within 10 years. But, and you might have noticed this, the furniture is the exact same quality. Going back to Hugh and the tarot-reader, "being the first in your family to die", sounds more awful than "not having to go to any family funerals". But they really mean the same thing.

This aspect is incredibly important for the business world. Imagine you work in a big company, and you're in charge of making a big decision. You get an email that is supposed to explain the pros and cons of making this decision, but instead you get an email explaining only the pros. If a rational person were to get that email, they would instantly ask for more information.

But, as you should know by now, there are very few rational people. In fact, most people would just read the email and be done with it.

How to avoid it

Going back to the example with the furniture, you can avoid making a bad choice if you look at it statistically. There might be a 90% guarantee that it will last 10 years, but that means there's a 10% chance it won't! If you're like Ike, you will make a nifty pie chart to avoid framing.

But, it's not that hard to calculate the percentages in your head.

Another thing you can do for this scenario: Take the information and reverse it so it still means the same thing. From there, you can look at both statements.

Of course, sometimes framing is just lack of information. In that case, do what Ike did with the email, and ask for the missing info.

How to exploit it

Now let's get a little bit devious. Hugh's son, Hector, wants to go with his friends to Ultimate Air Stadium, the trampoline park that just opened across the street, but he needs to get Hugh's permission first. And to make Hector's life harder, he just heard that 5% of people get injured at Ultimate Air Stadium. Luckily, Hector knows quite a bit about framing. Let's see what he says to Hugh, shall we?

And it works like magic.

Remember, though your parents know more than you about the world and you should always listen to them, they're not perfectly rational. Nobody is perfectly rational. So your parents will definitely fall for this.

Of course, there are more ways to exploit framing. Suppose you and your friends are going out to watch a movie, but you have to decide what movie to watch. You want to watch the action-adventure movie Dingo-Man, but one of your friends wants to watch the sci-fi movie Expedition 2479. It comes down to a vote, and both options are tied. The deciding vote is your older brother, who hasn't heard of either movie. You have to explain both movies to him. But, instead of giving a basic explanation for both, you use framing to your advantage.

When you explain the plot of Dingo-Man, stress on the good parts of the movie and make it sound better. For example, you could say: "The Oscar-winning thriller whose story of a half-man, half-dingo captivated audiences globally." But when you explain Expedition 2479, give a more basic explanation which doesn't make it sound as good as Dingo-Man. Like: "Some movie about a bunch of guys on a spaceship or something." Sure, you might not always get the chance to 'frame' a certain selection, but when you do, use framing wisely.

Chapter 5: Over your head

They actually progressed on their adventure during that break! Hugh just bought the Idol Of Endowment, which according to the map is crucial to finding the treasure!

Wait; hold on, is the idol saying something?

Apparently, they need to find a wise old man. Luckily, there are tons of old men walking around the carnival. We just need to figure out who is the wise one.

Wow, Hugh is so quick to jump to conclusions. He must have evidence that Ike doesn't have.

Looks wise? That's all he was going off of? Well, we might as well try.

Sigh. The guy who "looks wise" had no clue about the idol. Great job, Hugh.

That guy recognizes the idol! He must be the "wise old man"!

What?! Why did Hugh do that?

That's not a good explanation! Based on the actual evidence, that dirty guy is the most likely candidate for our "old wise man", yet Hugh dismissed him just because he didn't look smart!

But why does Hugh do this? Well, you should already know the answer by now; humans are irrational. More specifically, humans fall for the halo effect.

Before I start explaining, I'm going to bring you another example related to Donald Trump.

Before he was elected, Donald Trump had no political experience, in contrast to all the other candidates and well, every president ever. So why did people believe he could do so well as a world leader?

People thought that since he was so successful in business negotiations, he could also be successful in political negotiations. But here's the thing: Business negotiations and political negotiations are two very different things. Donald Trump created a halo for himself through his business, and that is what made people vote for him.

These halos can be very misleading. The "wise-looking" guy had no clue about the idol, but he had a halo over his head because of his sage-like appearance.

How to avoid it

First of all, the problem isn't in judging people. Judging people fairly is perfectly fine. The halo effect comes when you judge people unfairly; Hugh was biased against the wise old man because he was dirty.

So how do you fairly judge someone? Well, first you have to choose what you're judging. If you're deciding how smart somebody is, you're not going to judge their hygiene, because that has little to do with intelligence and is based on stereotyping. Instead, judge how they act, whether they seem to be doing smart things, et cetera.

Well Hugh, if you think that way, I'll make it easier for you!

You can avoid focusing too much on something irrelevant, if you look at multiple things when judging someone. If Hugh looked at the fact that the wise old man started approaching them when he saw the Idol, or that their other candidate had no clue about it, then he wouldn't have been so quick to judge.

How to exploit it

If you're feeling extra villainous, you can use halos to your advantage. By creating a halo for yourself, you'll make yourself look better in the eyes of most people. But how exactly do you create a halo?

First of all, you need to look at how one attribute "influences" another. For example, some people consider business and political skills to be intertwined, so a supposedly good businessman would be a good politician. Or perhaps your skill in one sport depends on your skill in a different sport, so a poor swimmer might have his biking skills underestimated.

Now that you know this, use it to your advantage. If you consider yourself a good biker, remind everybody of this and you might be considered for the swim team. If you're incredibly good at playing the cello, others might consider you incredibly good at playing the violin, if you remind them of your skill.

Chapter 6: The idol of endowment

Let's recap the story so far. The wise old man wasn't too upset, and is willing to lead Hugh and Ike to the temple. But he wants the Idol of Endowment in exchange. So Hugh gives him the idol, right?

What do you mean you're not giving him the idol? This is treasure beyond all wealth we're talking about! Probably more than the "outrageous" price you paid for the idol! Let's go back to when Hugh was buying the Idol.

See Hugh, you thought $100 was too expensive, and now you think that it's worth more than treasure beyond wealth! Why does Hugh put so much value on it now, and not when he bought it? Because of another bias: the endowment effect.

It might be, but idol or not, the endowment effect is a real bias. The basic idea is that people put more value on something when they own it. Hugh thought the idol was worth less than $100 before he owned it, but after he started thinking of it as his, he valued it over all the treasure he could find! Remember, what we're talking about is beyond all wealth, so if he thinks that idol is worth more, there must be something going on in his noggin. But behavioral economists proved that this is a real phenomenon. They brought a bunch of volunteers to a lab, and split them up into three groups: buyers, sellers, and choosers. The sellers were each given a nicely decorated mug, and told to name a price they would sell it at to the buyers, who had to choose what price they would pay for it. Interestingly enough, the sellers named much higher prices for the mug than the buyers. But it gets better. The third group, choosers, had to choose between the mug, and a certain amount of money which was worth as much as the mug. But the choosers' prices for the mugs were much cheaper than the seller's prices, even though they made the same decision! But they were different because the Sellers owned the mug, so they valued it more than the choosers and buyers, who didn't own it.

How to avoid it

One of the most foolproof ways to avoid the endowment effect is to get a second opinion. Say you're selling your old AlphaTron Zero game system, which you bought a few years ago for $400. It's still available in certain shops for around that price, so if you want people to buy it from you, you should sell it at a lower price, right?

Except you really won't. You'll consider the console to be worth more than what you bought it for, simply because you own it. Trust me, I have personal experience on this sort of thing. Even if you set a lower price, you still won't be able to negotiate with anybody who might buy it.

So, since you're biased, just get somebody else to tell you how much your AlphaTron should cost. Ask someone who doesn't own it, so that they won't be biased like you.

Chapter 7: No change to spare

Hugh finally gave up the idol to the wise old man, and he led them to the Cov'fefe Temple, which is supposedly where the treasure is hidden. But they have to cross a very large lake to get there. Luckily, someone left a motorboat on the bank of the lake.

However, the motorboat was set to medium speed before they got it, and they won't cross the lake fast enough with that pace. So shouldn't they set it to a faster speed?

No?! We need a faster speed if we're going to get across this endless lake.

Sigh. Here comes another bias. Hugh, like all other humans,

is inclined toward staying in the status quo. He's too paranoid to raise the speed, because he's allergic to change.

Oh, and he's also literally allergic to change. We're not talking about that right now, Hugh.

Oh, but this is a real problem. If you study physics, you might have heard about the law of inertia, which states that stationary objects will remain stationary unless an outside force acts on them. It's the same thing with people when making decisions. Humans won't look at all options equally when deciding. Instead, they'll want to stay with the status quo. Imagine you need to buy some coffee before work.

There are two coffee shops nearby, one of which you always go to. Both shops seem to have equal quality of coffee, but your regular store has slightly higher prices than the other store. Which store do you go to?

Your information is that they seem to have equal quality, but you choose to stick with your regular expensive coffee just to stay safe.

How to avoid it

The best way to avoid inertia is to be completely neutral. Let me explain: Going back to the coffee shop example, while you're choosing, you could act like there are two new coffee shops, and you have to choose one.

Well, that's a little extreme, but I'm sure it's not impossible to just act like you don't have a status quo option. In the case of the boat, you could decide the speed you want to go before you start.

But you don't even need to do all that really, because knowing the problem is almost as good as solving it.

Just like Hugh earlier, if you find yourself justifying an option because of inertia, you can catch yourself and think again.

How to exploit it

But then there are times when you can call the shots, when you can set the default options people use. In that case, you can use this bias to your favor.

The below graph shows the percentage of people in certain European countries willing to donate their organs.

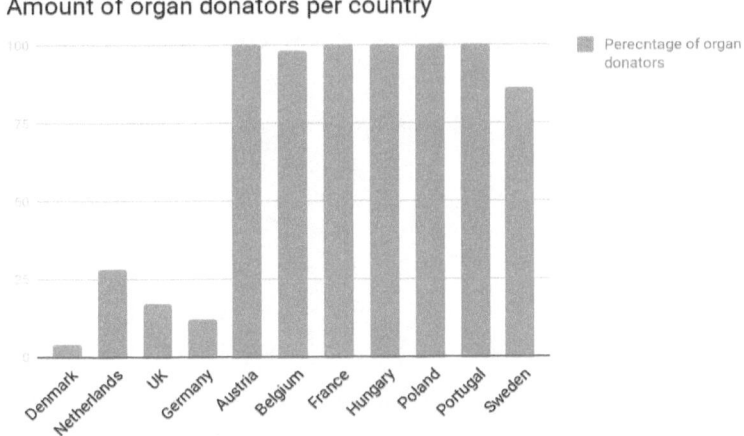

As you can see, the countries on the right have many more organ donors than the countries on the left. So what is the reason? Culture? Socioeconomics? Demographics? Consent form defaults?

Actually, "consent form defaults" is the correct answer. The countries on the left have a nifty little form which says, check this box if you want to participate in the organ donation program. But most people don't care what happens to their spleen after they're dead, so they just leave the box empty and don't donate their organs.

But then the countries on the right all have a nifty little form which says, check this box if you don't want to participate in the organ donation program. Most people don't care what happens to their spleen after they're dead, so they just leave the box empty, and donate their organs! It is amazing how it all works out in the end. Just like the person arranging the shelves in the anchor store had power over customers, the person designing that form can decide things as critical as somebody giving away their internal organs!

Chapter 8: Working it out

Where were we again? Oh, yes, Hugh and Ike managed to cross the lake despite Hugh's stubbornness. And Hugh and Ike entered the temple. Right now, they seem to be getting close to the treasure chamber, but it doesn't really matter, because they have to go through multiple grueling trials to get there.

That looks rough. In fact, Hugh is starting to get discouraged.

Oh no, they can't give up now! We still have many pages to get through! Time for market economics to save the day, Ike.

Smart move. If Hugh thinks about rewards while working, he should do better.

And it worked! In fact, Ike might know a bit more about humans than it may seem. But now our heroic duo reached another task.

This time it's a riddle. Hopefully Hugh and Ike can find the answer.

Time for market economics to save the…

Okay, it works for moving boulders, so I guess it should work for solving riddles.

Well, what's happening here? When Ike made Hugh think about the treasure, it made him lift the boulder, but distracted him from answering the riddle? Why are rewards a double-edged sword? Allow me to explain.

Most classic economists don't like math a lot. Which is why the rule of incentives is a very simple equation: More money equals more work. And this fact makes sense. Workers at an assembly line will work faster if rewards are higher. Hugh pushed harder when he thought about the treasure he was going to get. But does this approach work for mental tasks, like solving a riddle?

Incentives drive people to work because they make them focus on the task at hand. But for a cognitive task like solving a riddle, they distract you and stop you from thinking outside the box.

I have experimental data to back it up, too. Not from any other economist, but from a test using my own class as subjects. I told my entire class that I needed to do a survey for my research. But instead of a survey, I handed everybody a word search puzzle and told them to find as many words as possible, in the span of one minute. However, there were three different sheets. One sheet asked them to do the word search and didn't offer any sort of reward. Another sheet offered them a small Starburst candy as a reward for attempting the word search. And the third sheet offered them a large Mars bar for attempting the word search.

I collected the sheets (after giving students their rewards) and compared their offered reward to the number of words that they found. After I laid the data out, what I found was surprising. Those offered no reward did the best out of everyone on average. And those offered a high reward, the Mars bar, did the worst out of everyone on average! At first, I was completely baffled by this outcome. But then I realized that the incentives worked the opposite way; the incentives limited the students instead of driving them.

How to avoid it

So how can you be productive, and stop incentives from distracting you? Well, first of all, this might seem like terrible advice, but don't think about money.

Yes, I know, it's hard to do this, maybe even impossible. But at least try to clear your mind of all possible rewards.

Another good strategy is to try and make the task enjoyable. If you're doing something you're already passionate about, then just enjoy doing it instead of doing it for the reward. But if not, then try to have a pleasant working environment; challenge yourself to make work fun; or even think about what you're going to do after the task is completed (perhaps after finishing your math homework, you're going bowling with your friends).

How to exploit it

When you need someone to do cognitive work for you, like helping with your homework, big rewards will distract them and they'll be less productive. So, what do you do? One thing you can do is giving them the reward before the work, but threaten to take it away.

Let's say you're bribing your friend to help you with your homework. Give your friend $20, but say you'll take it back if he doesn't help you. That way, he won't focus on getting the reward, because he already has it. It'll also motivate him, because humans hate losing money more than they like gaining it. As we saw with the idol Of Endowment, people put more value on what they have than what they don't have.

In conclusion.... Wait, hold on...

Ike solved the riddle! And it's the final trial before the treasure chamber!

Now, all Hugh and Ike have to do is open that big chest!

Wait, the chest has a keyhole. The map didn't come with a key! Well this is it. I guess there's no hope now. We might as well find the boat somehow and...

Wow, that coffee cup Hugh was carrying with him all that time is the key to the treasure! The chest is opening!

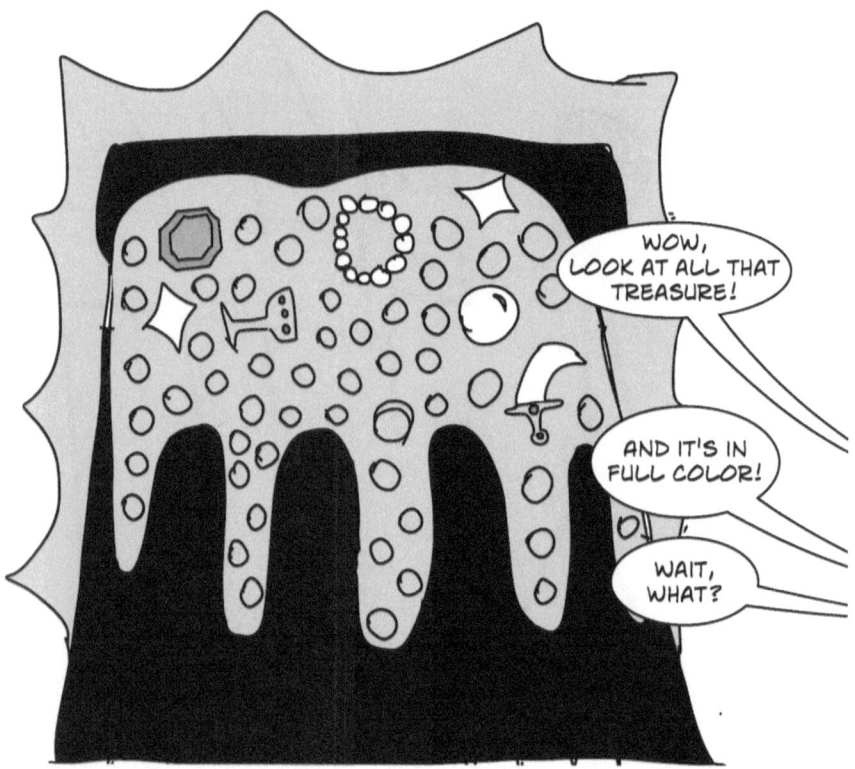

This really is treasure beyond wealth! Imagine what you could do with that! You could buy a palace, travel the world, you could afford an entire company...

What happened to the map! It's completely blank, except for some writing.

HAVE FUN AT THE BANK

Chapter 9: Fifty shades of green

You thought this story was finished? Well I said I would write a ten-chapter book, so here we are. And as you can see, Ike's in a bit of a predicament. So what is he doing anyway?

Well, leave it to Ike to do this kind of thing. He discovers a huge amount of treasure beyond any wealth in the world, and he decides to save it for his retirement! But this isn't a bad idea. Most people, humans included, save over $756,000 in some kind of retirement savings plan.

You can clearly see the contrast here. While Ike is putting his wealth in a savings account that will profit him in the long term, Hugh spent all his wealth on private jets, limousines, and gold watches. I bet he even bought that $480 anchor! And don't get me started on all the coffee.

In fact, just a few weeks ago, Hugh got a paycheck, and he put most of it in a pension plan! So why didn't he put any of his treasure in? Because not all money is created equal, and although all dollar bills might seem the same, they actually come in a whole wave of values. If you get some money every month from a office job, then you would spend it wisely. But if you suddenly get a huge sum of money all at once, whether it's through inheritance, winning the lottery, or in Hugh's case, discovering a millennium-old treasure hoard hidden by ancient economists, you'll feel the urge to spend it all.

Why does this happen? First of all, the amount of money you get is a huge factor. The treasure of Econ Island is literally beyond all wealth. This makes Hugh think that he can't run out of money, so he goes on a wild spending spree. But Hugh's paycheck doesn't have nearly as much money as the treasure, so he thinks about where to spend it.

There's also the factor of how hard it is to get money. Hugh and Ike both got a lot of money relatively easily.

Yes, I know it was really hard to get that treasure, but if you had to work an office job to get all that money in a short span of time, imagine how hard it would be!

And then there's mental accounting.

What I mean is, people designate certain amounts of money for certain things. For example, you would decide to spend $20 on a movie ticket. But this can have consequences. Say you bought the movie ticket, but on the way to the theater it got lost. Do you buy a new one?

Now imagine that same scenario, but this time you didn't buy the ticket before going to the theatre. And on the way to the theatre you lost a $20 bill. Do you buy a ticket and watch the movie?

Let's just take a moment to think about this. In each scenario, you lost a piece of paper worth $20. Except in the second scenario, that paper was green and had Andrew Jackson's face on it. How could that change anybody's decision? Except it does. In the first scenario, you already spent $20 on the movie, and that's all you were planning to spend. But in the second scenario, you didn't spend any money yet, so you were willing to buy a ticket. So what does this have to do with the treasure? Well, when you get money on a regular basis, you already have a plan to spend it on stuff like groceries and retirement savings. But if you just get a huge amount of money once, then you'll just spend it all quickly with no planning.

Chapter 10: Conclusion

So far, you've learned that humans make weird decisions left and right, including some which completely eschew any common sense or logic. And you might be wondering, why did I read this? Why did I read an entire book just to know that humans are stupid?

Well, now that you know where the human brain usually screws up, you can prevent yourself from making these absurd decisions. Not only that, you can also use these biases to your advantage. Just don't go too far with it, stay within the law, please.

So, use these tricks to your advantage throughout your life, and maybe read this book a second time, paying close attention to the 'how to avoid it' and 'how to exploit it' sections. And if your friends ever ask you how you're so much smarter than them, maybe recommend this book!

www.ingramcontent.com/pod-product-compliance
Lightning Source LLC
Chambersburg PA
CBHW031425210526
45464CB00005B/2058